# Capstone Design Courses: Producing Industry-Ready Biomedical Engineers

Capstone Design Courses: Producing Industry-Ready Biomedical Engineers

Jay R. Goldberg

ISBN: 978-3-031-00491-9        paperback
ISBN: 978-3-031-01619-6        ebook


DOI 10.1007/978-3-031-01619-6

A Publication in the Springer series
*SYNTHESIS LECTURES ON BIOMEDICAL ENGINEERING #15*

Lecture #15

Series Editor: John D. Enderle, University of Connecticut

**Series ISSN**
ISSN 1930-0328   print
ISSN 1930-0336   electronic

# Capstone Design Courses: Producing Industry-Ready Biomedical Engineers

Jay R. Goldberg, Ph.D., P.E.
Department of Biomedical Engineering
Marquette University

*SYNTHESIS LECTURES ON BIOMEDICAL ENGINEERING #15*

## ABSTRACT

The biomedical engineering senior capstone design course is probably the most important course taken by undergraduate biomedical engineering students. It provides them with the opportunity to apply what they have learned in previous years; develop their communication (written, oral, and graphical), interpersonal (teamwork, conflict management, and negotiation), project management, and design skills; and learn about the product development process. It also provides students with an understanding of the economic, financial, legal, and regulatory aspects of the design, development, and commercialization of medical technology.

The capstone design experience can change the way engineering students think about technology, society, themselves, and the world around them. It gives them a short preview of what it will be like to work as an engineer. It can make them aware of their potential to make a positive contribution to health care throughout the world and generate excitement for and pride in the engineering profession. Working on teams helps students develop an appreciation for the many ways team members, with different educational, political, ethnic, social, cultural, and religious backgrounds, look at problems. They learn to value diversity and become more willing to listen to different opinions and perspectives. Finally, they learn to value the contributions of nontechnical members of multidisciplinary project teams. Ideas for how to organize, structure, and manage a senior capstone design course for biomedical and other engineering students are presented here. These ideas will be helpful to faculty who are creating a new design course, expanding a current design program to more than the senior year, or just looking for some ideas for improving an existing course.

## KEYWORDS

capstone design courses, career preparation, biomedical engineering design education, design projects, new product development process.

# Contents

# Introduction

Two years after beginning my academic career as the director of the Healthcare Technologies Management program at Marquette University and the Medical College of Wisconsin, I became interested in teaching the two senior capstone design courses for biomedical engineers at Marquette University. As a biomedical engineer with 14 years of new-product development experience including 9 years of technical management experience with four medical device companies (DePuy, Baxter, Surgitek, and Milestone Scientific), I had a unique appreciation for the value of these design courses. My industry perspective was helpful in revising BIEN 146 and 147, *Principles of Design* and *Senior Design*, respectively, with my colleagues Dr. Martin Seitz (electrical and computer engineering) and Dr. Joseph Schimmels (mechanical engineering) to make them more relevant and better prepare our biomedical engineering students for careers in biomedical engineering.

In my opinion, senior capstone design courses are the most important courses our engineering students will take in their undergraduate programs. They provide students with an opportunity to apply what they have learned in previous years and develop their communication (written, oral, and graphical), interpersonal (teamwork, conflict management, and negotiation), project management, and design skills. These skills are needed for any career path, whether it leads to the medical device industry, medicine, law, business, or other area. Working in teams to identify problems and design, develop, and implement solutions to these problems is the best preparation for a career in biomedical engineering.

During each of the 7 years that I taught these courses, I solicited student feedback to determine how well the courses met our desired learning objectives and how the courses could be improved. I attended American Society for Engineering Education, Biomedical Engineering Society, and Biomedical Engineering Innovation, Design, and Entrepreneurship Alliance meetings as well as the National Capstone Design Course Conference and other meetings to learn and share best practices in the area of biomedical engineering design education and senior capstone design course management. This includes course administration issues such as lecture topics, assessment tools (examinations, peer evaluations, etc.), methods of team formation, types of design projects, course deliverables, faculty involvement, organizational support, and solicitation, funding, and industry sponsorship of projects.

This book provides ideas for how to organize, structure, and manage a senior capstone design course for biomedical and other engineering students. It will be helpful to faculty who are creating a new design course, expanding a current design program to more than the senior year, or just looking for some ideas for improving an existing course. Much of what is presented here has been published in a column (*Senior Design, IEEE Engineering in Biology and Medicine Society Magazine*) that I have written for the last 4 years. Some of the ideas presented reflect methods, processes, and components that I have included in my course at Marquette University. Other ideas have been successfully used in courses taught in other engineering programs.

Not all capstone design courses, biomedical engineering programs, and curricula are the same. Some programs emphasize preparation for employment in industry; others may focus on preparing graduates for medical school or academia. Each program has its own culture, values, and attitudes toward engineering design, which affect the level of importance and priority assigned to the senior capstone design course. Some curricula allow room for a two-semester design course sequence; others may only have room for one course. What works well for one program may not work well for others.

I hope you find this book useful in designing, organizing, managing, and improving your senior capstone design course. The better we can make these courses, the better prepared our students will be for meaningful, successful careers in biomedical engineering.

• • • •

Jay R. Goldberg, Ph.D., P.E.
Department of Biomedical Engineering
Marquette University
jay.goldberg@mu.edu

# SECTION I

## Purpose, Goals, and Benefits

CHAPTER 1

# Why Our Students Need a Senior Capstone Design Course

Several industry surveys and academic studies confirm that graduating engineers are inadequately prepared for careers in industry.[1–5] Among the competency gaps identified in these studies were teamwork, communication, business skills/knowledge, and an ability to interact with diverse multidisciplinary groups. Employers want to hire engineering graduates who can communicate well with a variety of diverse personnel and work effectively as members of project teams. There is a common feeling that "while engineering graduates are technically savvy, sometimes they lack training in the science of soft skills."[6] To meet the educational needs of engineers and their employers, engineering schools throughout the United States need to produce graduates with not only strong technical skills, but also the ability "to work as part of teams, communicate well, and understand the economic, social, environmental, and international context of their professional activity."[2]

In the United States, there are more than 50 undergraduate biomedical engineering programs that include a senior capstone design course in their curricula. These courses are the culmination of a student's first 3 to 4 years of their undergraduate engineering education. They provide students with an opportunity to work in teams and apply what they have learned in previous course work to the solution of an open-ended real-world problem. Appropriately structured senior design courses also provide students with opportunities to develop their design, analytical, project management, communication (written and oral), and interpersonal (teamwork, negotiation, and conflict resolution) skills. They can also provide students with an understanding of the economic, financial, legal, and regulatory aspects of the design, development, and commercialization of medical technology. Many senior design courses provide exposure to the medical device industry through industry sponsorship of senior design projects. A senior design course can be an excellent vehicle for preparing students for careers in biomedical engineering, and is perhaps the most important course that our students will take as undergraduates.

Senior capstone design courses possess a unique potential to prepare students for careers in biomedical engineering. Whether students work in the medical device industry, a hospital, or consulting firm or attend medical, dental, law or graduate school, the knowledge gained and skills developed in these courses (if properly structured) will be helpful in their careers. The real-world problem solving team experiences that these courses provide benefit all students.

## REFERENCES

1. American Society for Engineering Education, "Summary report on evaluation of engineering education," *Journal of Engineering Education*, pp. 25–60, September 1995.

2. *The Green Report—Engineering Education for a Changing World*, American Society for Engineering Education, Washington, DC, 1994.

3. Cherrington, B., et al, *The Engineering Leader and Leading Change: A Report from the ASEM Team*, presented at the 1995 National Conference of the American Society for Engineering Management.

4. *Manufacturing Education Plan (MEP)*, Society of Manufacturing Engineers, Dearborn, MI, 1998.

5. Bahner, B., "Report: Curricula need product realization," *ASME News*, vol. 15, pp. 1–6, March 1996.

6. O'Shei, T., "Engineers are taught how to be true team players," *Business First*, 1998, http://buffalo .bizjournals.com/buffalo/stories/1998/07/20/focus6.html.

• • • •

CHAPTER 2

# Desired Learning Outcomes

Should the goal of senior design courses be the development of design skills, career preparation, or both? After graduation, most graduates of many biomedical engineering programs find jobs as engineers in the medical device industry. Some will attend professional school (medical, law, dental, or business) or graduate school. Many of those with advanced degrees in engineering will eventually be employed in the medical device industry. If career preparation is a goal of senior design courses, should the focus be on industry, where most biomedical engineering graduates begin their careers, or should it include alternate career paths? What should be the desired outcomes of senior design courses?

At Marquette University, the senior capstone design courses for biomedical, electrical, computer, and mechanical engineering students are taught together. Project teams consist of three to five students from different engineering disciplines and are advised by faculty members and industry sponsors. The design project is the focus of the course, and students are required to hand in specific project-related deliverables during the two-semester course sequence. Twice a week, students from all engineering disciplines attend common lectures on various topics related to their projects, the product development process, or their careers. Biomedical engineering students meet separately six times each semester to discuss issues specific to the field.

The desired outcomes of the course at Marquette University include the following:

- development of design and technical skills
- development of "soft skills" such as teamwork, communication, and interpersonal skills
- preparation for careers in biomedical engineering (with emphasis on careers in industry)

To achieve these outcomes, the primary objectives of the course include development of the following:

- design skills
- project management skills
- the ability to manage the product development process
- the ability to work effectively in teams
- oral, written, and graphical communication skills

The secondary objectives of the course include the following:

- experience with solving a real-life, open-ended problem
- development of an understanding of the industry perspective (including financial, regulatory, and legal issues)
- exposure to medical and surgical procedures and technologies
- exposure to results-oriented evaluations of their projects

The course syllabus includes lecture topics in three areas. The first group of topics covers topics important to the successful completion of the projects. The second group of topics covers issues important for the students' careers and may be applicable to their projects. The third group of lecture topics includes issues of specific interest to biomedical engineers.

Students are provided with opportunities to develop their communication skills through interaction with other team members, required oral project status updates, written proposals, and final reports, and various written deliverables throughout the course. The philosophy of the course is that a team-based project experience requiring specific project deliverables (similar to those required in industry), supplemented with a series of project and career-related lectures will prepare students for careers in industry. This approach can fill some of the competency gaps often cited in the skill sets of new engineering graduates.

Although the focus of the course is on preparation for careers in industry, biomedical engineering students seeking careers in medicine will benefit from the skills developed and lectures presented in this course. The ability to work in teams, good oral and written communication skills, knowledge in conflict resolution, familiarity with career management issues, and an understanding of the economic, legal, and regulatory aspects of health care delivery and medical technology development will be extremely helpful to these students when they manage their own medical practices, interview for jobs, or work as consultants to medical device companies.

• • • •

CHAPTER 3

# Changing Student Attitudes, Perceptions, and Awareness

Senior capstone design courses provide an excellent opportunity for biomedical engineering design instructors to help students begin to look at the "real world" differently. Depending upon the structure of a particular institution's senior design course and the resulting design experience obtained, students' attitudes and perceptions related to biomedical engineering design, the product development process, and the engineering profession can be positively affected. An increased awareness of employer expectations, accepted practices, and the constraints of medical device design can result in new ways of thinking that will help prepare students for careers in biomedical engineering.

Components of the senior capstone design course experience often require students to think about and do things differently than they did before taking the course. They can transform the way students think about the design process and how it is implemented, their roles on a project team and within an engineering organization, and how they will be expected to perform their jobs. These components include learning to solve open-ended design problems, developing a revised definition of design, realizing the importance of teamwork, acknowledging the need for a variety of skills for successful engineering careers, and experiencing performance evaluations based on project outcomes and team performance.

In many engineering courses, students solve problems that have one correct answer. They learn to apply formulas and equations to calculate answers to problems presented in engineering texts. Often, the correct answers are found in the back of the textbook to instantly confirm the correct solution to the problem. Once the correct answer is found, the problem is solved, and no additional work is required. Senior design projects require a different approach to problem solving. Reflective of the real world of engineering design, these projects present open-ended design problems that have no singular "final" solution. There are many ways to solve the problems presented in a senior design project. Students eventually learn that there is no singular correct solution to a design problem and that any design that meets customer needs and fits within the technical, economic, legal, and regulatory design constraints is an acceptable solution and, therefore, an acceptable endpoint for the project.

For many students, the definition of a good design is simply one that meets all technical and functional performance requirements. Senior design projects require students to revise and expand this definition. Students learn that customer needs are not limited to technical and functional performance requirements but include additional requirements that allow the design to fit within all of the constraints of medical device design. Their expanded definition of a good design is one that meets not only all technical and functional performance requirements but all financial, marketing, safety, legal, and regulatory requirements.

Some biomedical engineering programs provide an opportunity for students to work on group projects in courses throughout the undergraduate curriculum. For many students, the senior design course may be their first team project experience. Up until this point, grades were based on individual performance. This gave the student some degree of control over the grade they received on individual assignments and projects. Each student completed all tasks required by the assignment or project and had control over the quality of their own effort. If a student wanted a high grade, he or she simply invested more time and effort in the project. Senior design projects, typically completed by teams, do not give individual students as much control over the quality of the final design. An individual student has control over the quality of their own output but not that of other team members and is not responsible for nor will complete all tasks required by the project. Work is distributed among the team members who, because of different experiences, perspectives, attitudes, and opinions, may favor different design approaches and strategies. All team members may not invest similar amounts of time and energy into completing their assigned tasks. These differences in individual attitudes and performance often lead to conflict within the team. Students quickly learn that their individual success in the course is closely tied to the performance of their team members, not just their own individual performance as with other courses. This teaches students to learn to work together as a team, help each other when necessary, and recognize other viewpoints and opinions. When members of project teams do not share the same goals for project and course grades or are not "pulling their weight" on the project, students desiring higher grades will need to employ their conflict management and negotiation skills to convince team members to invest more time and energy in their respective assigned tasks. If this fails, these students may need to take on more than their fair share of the workload to improve the team's final project grade. These situations are representative of the experiences they may encounter as members of project teams in industry and help prepare students to deal with team-related issues.

Many engineering students believe that design skills are the most important skill set for engineers to possess. There is a saying among some engineering managers in industry that "Engineers are hired for their technical skills, but fired for their [lack of] communication skills. Design skills are important, but communication and interpersonal skills are equally (and in some cases more) important to career success. The team experiences obtained through completion of senior design projects help students realize the importance of written and verbal communication, negotiation, conflict management, and project management skills to successful careers in engineering as well as medicine, law, dentistry, business, and other career paths open to biomedical engineering students.

In many undergraduate courses, level of effort is often a significant component of a student's final grade. If not all course objectives were met by a student but acceptable mastery of the subject matter and a high level of effort were demonstrated, the student's grade may be elevated to that of another student who met all objectives of the course. In industry, performance evaluations are based almost completely on objectives, many of which cannot be met without the help of the project team. If an engineer meets all of the established objectives for the year, an appropriate increase in salary will typically result. Exceeding one's objectives warrants a higher salary increase. Not meeting all objectives results in a nominal raise, no raise, or sometimes termination of employment (if performance is chronically at this level). A high level of effort is appreciated, but if not accompanied by improved outcomes, will often not improve a performance evaluation. The senior design course can be used to introduce students to the outcomes-based evaluation and grading method commonly used in industry. For example, at Marquette University, students in the senior design course receive a score of 85 (B) if all minimum requirements have been met for a particular deliverable. Grades above 85 are indicated only if the team exceeds the minimum established requirements, and grades below 85 are given for work that does not meet minimum requirements. This outcomes-based grading system is heavily based on team performance and reflects the industry model. However, because the objectives of the course are more academic, emphasis is placed on learning about the product development process and project management and developing communication and interpersonal skills. The final course grade considers individual performance and level of effort and the team's project grade, which is based on project outcomes.

The senior design experience provides opportunities for course instructors to transform the way students think about the design process, teamwork, expected job performance, and the engineering profession. The resulting changes in attitudes, perceptions, and awareness can play an important role in preparing students for careers in biomedical engineering.

•   •   •   •

CHAPTER 4

# Senior Capstone Design Courses and Accreditation Board for Engineering and Technology Outcomes

As a condition for accreditation, the Accreditation Board for Engineering and Technology (ABET) requires all undergraduate engineering programs in the United States to demonstrate that their programs produce eleven specific learning outcomes (ABET criterion 3 outcomes).[1] These outcomes are specific abilities, knowledge areas, and attitudes that all students should possess upon completion of the undergraduate engineering program. ABET reviewers are looking for results of self-evaluations and assessments that prove that the required outcomes are being produced.

Biomedical and other engineering programs conduct reviews to determine which outcomes are met by the courses in their respective curricula. In situations where a specific outcome is not produced, programs are required to develop and implement plans for improvement to ensure that all requirements will be met. These plans may include development of new courses or modifications to existing courses. Programs must also document changes and eventually show that the changes resulted in improvements. The cycle of assessment, gap analysis, and change implementation closes the feedback loop and is very similar to the process required of companies by the ISO 9000 family of standards.

According to the *2006–2007 Criteria for Accrediting Engineering Programs*,[1] Engineering programs must demonstrate that their students attain

- an ability to apply knowledge of mathematics, science, and engineering;
- an ability to design and conduct experiments, as well as to analyze and interpret data;
- an ability to design a system, component, or process to meet desired needs within realistic constraints such as economic, environmental, social, political, ethical, health and safety, manufacturability, and sustainability;
- an ability to function on multidisciplinary teams;
- an ability to identify, formulate, and solve engineering problems;

- an understanding of professional and ethical responsibility;
- an ability to communicate effectively;
- the broad education necessary to understand the impact of engineering solutions in a global, economic, environmental, and societal context;
- a recognition of the need for, and an ability to engage in lifelong learning;
- a knowledge of contemporary issues;
- an ability to use the techniques, skills, and modern engineering tools necessary for engineering practice.

Biomedical engineering programs must also "demonstrate that graduates have an understanding of biology and physiology, and the capability to apply advanced mathematics (including differential equations and statistics), science, and engineering to solve the problems at the interface of engineering and biology; the ability to make measurements on and interpret data from living systems, addressing the problems associated with the interaction between living and non-living materials and systems."[1]

Many capstone senior design courses include lectures to develop students' knowledge of the product development process, project management, professional engineering practice, and the regulatory, legal, ethical, and economic aspects of medical device design. They also provide students with the opportunity to develop design, communication, and interpersonal skills through a team-based project experience. Many of the ABET criterion 3 learning outcomes focus on the development of the same knowledge areas and skill sets emphasized in senior capstone design courses. Thus, these courses can play an important role in helping undergraduate engineering programs meet many of the ABET learning outcome requirements.

A thorough assessment of a well-designed senior capstone design course can indicate to what degree the course can assist the program's efforts to meet the requirements. For example, a recent self-assessment of the capstone senior biomedical engineering design courses at Marquette University was conducted by the biomedical engineering faculty. The faculty developed a list of performance criteria that could be used to indicate that a specific learning outcome was being produced (performance indicators). They also developed a list of assessment tools such as examination questions, final reports, oral presentations, and other course deliverables that could be used to demonstrate that performance criteria were met. For example, outcome C (ability to design a system, component, or process to meet desired needs within realistic constraints such as economic, environmental, social, political, ethical, health and safety, manufacturability, and sustainability) was assessed using the following performance indicators:

- defines customer needs
- defines design constraints

- offers alternative solutions
- defines problems to be solved
- defines project scope
- compares alternative solutions
- defends selection of final design
- build prototype to meet needs
- validates performance of prototype

These performance indicators were evaluated using the following assessment tools (team-written documents that are required deliverables of the courses):

- project definition document: contains project objective statement (which defines problem and project scope), existing solutions, and design constraints
- customer needs document: contains list of customer needs along with design constraints
- generated concepts document: contains potential solutions generated by project team
- final concept document: defends selection of proposed final design
- experimental validation document: contains test protocols, test results, data analysis, and conclusions regarding how well prototype meets performance requirements
- final report: contains final design, test results, information regarding how well customer needs were met
- prototype

The results of the assessment indicated that upon completion of the two-course capstone senior design sequence, most students were demonstrating the abilities, attitudes, and mastery of knowledge required by ABET learning outcomes A–K. There were components of the two courses that contributed (to different degrees) to the production of each of the learning outcomes. For example, because of the project team experience, the course played an important role in producing outcome D (ability to function on multidisciplinary teams). However, because of the structure of the course, it played a negligible role in producing outcome I (recognition of the need for, and an ability to engage in lifelong learning). Each course in the curriculum contributed to the production of some of the learning outcomes. However, when assessed along with other courses in the curriculum, the program was shown to produce all of the ABET learning outcomes.

Senior capstone engineering design courses typically include a wide variety of lecture topics and provide students with many opportunities to develop design, communication, and interpersonal skills. This learning environment can play an important role in producing the desired ABET learning outcomes. Careful identification and assessment of appropriate performance indicators using the appropriate assessment tools can help a biomedical engineering program determine the role of their senior capstone design course in producing the desired ABET learning outcomes.

## REFERENCE

1. *2006–2007 Criteria for Accrediting Engineering Programs*, ABET, Baltimore, MD, www .abet.org.

•  •  •  •

# SECTION II

# Designing a Course to Meet Student Needs

CHAPTER 5

# Course Management and Required Deliverables

Almost all senior capstone design courses have similar goals but different ways of achieving them. There is no one correct way to structure a capstone design course. The optimum structure depends upon the following:

- culture and focus of the particular biomedical engineering program (emphasis on preparing students for industry, medical school, or academia)
- curriculum (credits available for the capstone design course)
- availability of internal resources such as faculty advisors, a prototype shop, laboratories and test equipment, and funds for projects and guest speakers
- availability of external resources such as medical or dental school faculty, industrial design students, business students, industry sponsors, supportive alumni, and guest speakers

In 2005, two studies were conducted to collect survey responses from 444 engineering (not limited to biomedical engineering) programs associated with 232 institutions.[1,2] The results illustrate the different ways senior capstone design courses are managed and structured.

## 5.1    DURATION OF CAPSTONE COURSE

Most courses were either one (47%) or two (32%) semesters long. Some were as short as a few weeks and others as long as three or more semesters.

## 5.2    STRUCTURE AND SEQUENCE OF CAPSTONE COURSE

Most courses (55%) involved attending classes while working on the project. Some courses (22%) required classwork before working on the project. Other courses (21%) required the project only, with no required classes. Two percent of the courses involved only classes with no project.

## 5.3    FACULTY INVOLVEMENT

In 16% of engineering departments responding to the survey, 100% of the faculty was involved in the course. In 20% of the programs, at least 80% of the faculty was involved in the course. At the other end, 40% of programs involved 20% or less of their faculty. Faculty members acted as project mentors (57%), consultants (34%), and evaluators (16%).

## 5.4    GRADING

In capstone design courses, the emphasis on project work results in a significant portion of a student's grade being dependent upon team results, not just individual student performance. Thus, the final course grade is often a combination of team and individual performance. Tools to assess individual performance include peer reviews and individual deliverables such as quizzes, examinations, and other assignments. Team performance can be assessed via team deliverables including the final report. Per the 2005 survey, individual final course grades were determined through evaluation of individual (53% of respondents) and group (67%) deliverables, the final report or other final group deliverable (86%), team peer evaluations (57%), and other deliverables (31%).[2]

## 5.5    REQUIRED COURSE DELIVERABLES

Many capstone design courses require written documents, individual and team oral presentations, maintenance of a team project notebook, functional final prototypes, poster presentations, and other assignments. Many of these deliverables are similar to those that students will create if employed in industry. Written deliverables may include documentation of the project objectives, customer needs, target specifications, generated concepts, and final concept(s). In addition, students are often required to generate a project schedule and risk management plan, manufacturing document, experimental validation document, and final report. Oral presentations may include periodic project status updates (similar to what they would provide to managers and peers in industry) and final proposals and/or reports. In many courses, students are expected to develop a final functional prototype and provide test data to prove that all required performance specifications and customer needs have been met. Some courses require students to conduct patent and standards searches, write environmental impact statements, estimate product costs, and formulate strategies for regulatory approval. Many programs require students to create poster exhibits for design competitions that provide project teams with the opportunity to showcase their design projects to the university and local communities. Programs that include entrepreneurship as an important component of their capstone design course often require teams to write business plans.

## 5.6    EXTRACURRICULAR ACTIVITIES

Depending upon the proximity of an engineering school to medical device companies, hospitals, and clinics, capstone design courses can provide students with the opportunity to tour manufacturing facilities, observe surgical and medical procedures, and interact with patients and medical personnel. These experiences can be very helpful to students trying to identify problems and opportunities on which to base their design projects.

## REFERENCES

1. Howe, S., and Wilbarger, J., *2005 National Survey of Engineering Capstone Design Courses*, presented at the 2006 ASEE Annual Conference and Exposition, Chicago, IL, June 2006.
2. Wilbarger, J., and Howe, S., *Current Practices in Engineering Capstone Education: Further Results From a 2005 Nationwide Survey*, presented at the ASEE/IEEE Frontiers in Education Conference, San Diego, CA, October 2006.

•   •   •   •

CHAPTER 6

# Projects and Project Teams

The most important component of a senior capstone design course is the design project. Working on projects allows students to develop their project management skills while learning about design and the product development process. Working on self-managed project teams allows students to develop their teamwork, negotiation, conflict management, interpersonal, communication, and time management skills. To best simulate the environment in which students will be working after graduation, project teams should consist of members who collectively possess the variety of skills needed to successfully complete their projects. In many biomedical engineering capstone design courses, teams consist solely of biomedical engineering students. In other courses, teams may consist of a mix of biomedical, mechanical, electrical, or other engineering students. Some courses include students from nonengineering disciplines such as business on their project teams. To create truly multidisciplinary teams, business, law, or science students could be part of the team along with physical therapy, nursing, dental, or medical students. The addition of industrial design students would further enhance the project team. Members of multidisciplinary project teams provide the required expertise for the project, teach each other about their respective disciplines, learn to respect each other's opinions and perspectives, and appreciate each member's unique contribution to the project.

Results from the 2005 survey of capstone design courses indicate that 18% of respondents reported using project teams consisting of one student, 81% used teams consisting of students from one discipline, and 35% included teams consisting of students from more than one department.[1] Students from various engineering departments participated in 76% of interdepartmental capstone projects. Business students participated in 4% of the projects. Twenty percent of the projects included students from other departments such as physics and theater.

To optimize a student's experience with the senior capstone design course, it is important to match his or her interests and career goals with the objectives of and learning opportunities provided by the project. In forming project teams, team members' interests, previous work experience, and skills should be considered. Ideally, teams should consist of students who possess the required characteristics and skills for a particular project and rank the project among their top choices. If possible, teams consisting of members with complementary personality types (assessed by a Myers–Briggs or similar test) should be formed. This can help prevent one team from having two strong leaders who constantly clash or struggle for power or too many followers waiting for a team leader to assign them work.

Team size is an important consideration. Individual students working on a project by themselves will not benefit from the team experience nor develop their team skills. On a team consisting of two students, if one does not complete his or her work assignments, the other may have to carry a heavier work load. Three to five students can be an optimum project team size depending upon the complexity of the project. In teams with six or more students, there might not be enough challenging work to assign to all team members, enabling less productive and motivated students to pass the course without completing their share of the work. In complex projects, a team of six or more students could be divided into smaller teams, each focusing on a particular subproblem or subsystem of the final design. In the 2005 survey of capstone courses, 2% of respondents reported using only teams comprised of one student. Average team sizes of 1 to 3 students (30%), 4 to 6 students (60%), 7 to 9 students (3%), and 10 or more students (4%) were also reported.[1]

There are several types of design projects often found in the project portfolios of capstone design courses. The goal of each is to develop a design solution to a specific problem. Each type provides a different learning experience for the students. Some may emphasize clinical interaction, market analysis, business planning, or cultural and political issues more than others.

Biomedical engineering design projects are typically sponsored by faculty members, medical device companies, or government agencies. Some project ideas are generated by students who wish to work on projects of interest to their co-op or internship employers, continuing work they started while working for these employers or starting a completely new project. Other project ideas often come from entrepreneurial students who are interested in eventual commercialization of their final products. Respondents to the 2005 survey of capstone design courses reported receiving project ideas from industry (71%), faculty research (46%), external design competitions (24%), students (15%), books (6%), and other sources (21%).[1]

Ideally, students should be given the opportunity to observe the use of medical devices in the clinical setting and encouraged to identify problems that could then be the focus of their capstone design projects. Students would learn how to observe, question, listen, and interact with medical personnel to identify problems and associated opportunities for commercialization.

Faculty-sponsored projects are typically funded by a faculty member, either through a grant or other source, and often focus on the development of a tool, apparatus, or piece of equipment to be used to conduct research in the laboratory of the faculty sponsor. To provide an adequate design experience to students, these projects must contain a significant design component, including identification of customer needs, development of potential designs, and construction and testing of prototypes. They should not require students to conduct research for the faculty sponsor and write research papers for publication.

Medical device companies often sponsor projects to develop solutions to defined problems of interest to the company. Although these projects do not provide opportunities for teams to find problems on their own, they provide many other advantages to students and the sponsoring companies. Students get the opportunity to work on real-world problems of importance to industry and learn about the needs of the medical device market and the operations of a company. They learn firsthand about the requirements and constraints of medical device design. Experience gained from industry-sponsored projects helps prepare students for careers in the medical device industry.

Companies sponsoring design projects benefit by 1) gaining additional technical resources, 2) involvement and participation in the training of new engineers, and 3) advertisement of the sponsoring company on campus. Sponsorship of a project provides a company with a team of engineering students dedicated to working on a solution to the company's problem, at no labor cost to the company. This can be very beneficial to companies with limited engineering resources by allowing students to work on lower priority projects while the company's resources are directed toward work on higher priority projects. Sponsorship provides companies with access to and a higher profile among graduating engineers, which can assist in recruitment efforts.

Student-generated projects provide them with the opportunity to identify a problem of interest to the team. This can lead to a higher degree of buy-in among team members (compared with team members who are assigned to projects that they may or may not be interested in), resulting in a highly motivated and enthusiastic team. It can be helpful to solicit project ideas from all upcoming senior design students at the beginning of the summer to give them time to investigate areas of interest and develop their own ideas for potential projects. The summer months provides time for ideas to incubate so that students are ready to form teams and begin work on their projects at the start of the fall semester. Many of the students working on these projects are interested in entrepreneurship. Their interest in potential commercialization often motivates them to delve deeper into the financial, regulatory, and intellectual property aspects of their projects. Student-generated projects do not have the financial resources available to those sponsored by industry or faculty. Department or college funds should be made available to these project teams to provide the resources needed for prototypes and other supplies.

Projects sponsored by industry, faculty, and students involve different objectives and expectations of students and may emphasize different components of the design process. They can be further classified according to the end user, target market, or stakeholders of the project:

1. Industry-sponsored projects
   Emphasis:
   - lower costs / higher profits

- ease of use, increased safety, reduced liability
- time savings
- increased quality of health care

Examples:
- development of new device or method
- improvement to existing product or method
- cost reduction
- improved packaging or production process

2. Student-sponsored entrepreneurial projects
Emphasis (similar to industry-sponsored projects):
- emphasis on newer and better product for commercialization
- students often develop business plans including market studies, plans for intellectual property protection, and regulatory strategies

3. Service learning projects
Emphasis:
- improving access to health care for underserved populations and developing nations
- development of low-cost affordable medical technology
- typically not for profit

Examples:
- development of an affordable, lower cost design alternative to an existing device or procedure
- easier and simpler manufacturing or assembly process to allow local assembly of new device with locally available materials

4. Assistive technology projects
Emphasis:
- projects to benefit people with disabilities
- typically designed for a specific patient (custom design) with a specific disability or condition
- lower cost, more affordable than existing technology

Examples:
- design of new assistive device
- custom design modification to existing assistive device to accommodate a specific patient
- improvement to existing product or method (increased usability, safety)
- cost reduction to make assistive technology more affordable and accessible

The design project is the most important component of the senior capstone design course. To best prepare students for careers in biomedical engineering, they should be given the opportunity to work on projects with teams that simulate the team environment in an industrial setting. Team composition is important in providing as much of a multidisciplinary team experience as possible. The type of project to which students are assigned should be based on the student's interests and career goals and the experience and skills they bring to the project.

## REFERENCE

1.  Howe, S., and Wilbarger, J., *2005 National Survey of Engineering Capstone Design Courses*, presented at the 2006 ASEE Annual Conference and Exposition, Chicago, IL, June 2006.

•   •   •   •

CHAPTER 7

# Lecture Topics

Although the design project should be the focus of the senior capstone design course, lectures are an important component of the course. They provide an opportunity to present information needed by students to properly execute their projects and/or prepare them for their careers and can be used to supplement reading assignments or introduce new material not presented elsewhere in the course or curriculum. In courses that include students from more than one engineering discipline, topics of interest to all disciplines can be presented in joint lectures and topics of importance to biomedical engineers can be presented in separate discipline-specific lectures.

According to the 2005 survey of capstone design courses, 35% of respondents taught one course in which students from all engineering disciplines were presented with the same lecture topics.[1] Separate courses for students from each discipline including different topics were reported by 24% of the respondents. Students from different engineering disciplines shared the course but received both common and discipline-specific instruction in 11% of courses. In 20% of the courses, no classroom instruction or lecture topics were included.

Team-based project experiences help students develop their communication, conflict management, negotiation, and teamwork skills. Lectures on these topics can be used to make students aware of problems that often arise when working on teams and show them how to solve these problems through role playing or classroom discussions. Ways of dealing with commonly observed team-related problems such as 1) students insisting that their ideas are always the best and refusing to listen to other ideas, 2) students failing to attend team meetings, and 3) students not completing their share of the project workload can be presented through reading assignments and/or lectures. This will prepare students to deal with these and similar problems when they occur on their capstone project teams and on their industry project teams.

Some engineering programs require a course that deals with professional practice and career management issues. For those programs that do not require such a course, the addition of lectures that present topics such as

- what to consider when searching for a job,
- how to evaluate a job opportunity,

- salary negotiations,
- what to focus on during the first year of the job,
- how to get promoted,
- how to conduct a job search while working,
- how to leave a job,
- and professional development opportunities

are very helpful to students as they prepare to leave the academic setting and begin their careers after graduation.

Class time is not unlimited. Lecture topics should be chosen carefully to make the best use of class time. If a topic under consideration is needed by students to 1) effectively manage their projects or 2) prepare them for work as professional engineers, then the topic should receive a high priority. Lecture topics that can help students manage their projects or prepare them for work as engineers (or both) include the following:

- Project management
    - project kickoff and update meetings
    - scheduling methods
    - work breakdown structures
    - project risk management
    - preventive actions and contingency plans
    - development planning and design reviews
- Teamwork
    - negotiation skills
    - conflict management
    - team building
    - dealing with team problems
- Communication skills
    - oral presentations
    - written and graphical communication
- Product development/design process
    - project definition
    - customer needs identification
    - target specifications
    - generated concepts
    - final concepts
    - prototyping

- testing for safety and efficacy
- production
- commercialization
- Design issues
  - risk management
  - safety
  - human factors/user-centered design
  - industrial design
  - packaging and sterilization
  - environmental impact (energy efficiency, sustainability)
  - prototyping methods
- Constraints of medical device design
  - economic (competition, insurance reimbursement, cost containment, etc.)
  - regulatory (clinical studies, design controls, quality systems, pathways to market, product recalls, complaint investigations)
  - legal (intellectual property, search for prior art, liability)
  - ethical (informed consent, patient privacy, use of animals)
  - compliance with industry standards (ISO 9000, ASTM, AAMI, RESNA, IEC, etc.)
  - political, social, and cultural issues
- Business
  - medical device industry
  - economic analysis
  - finance
  - marketing
  - globalization
  - project portfolios
- Entrepreneurship
  - identification of opportunities for commercialization
  - sources of funding
  - business plans
- Career management
  - career opportunities for biomedical engineers
  - considerations for first job
  - salary negotiations
  - goals for the first year of the first job

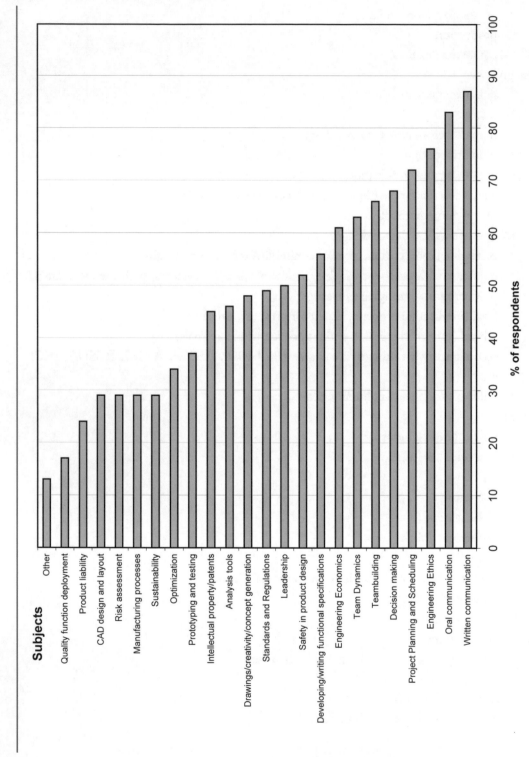

**FIGURE 7.1:** Subjects taught in senior capstone design courses.

- ○  career advancement
- ○  professional development
- Other
  - ○  personal finance
  - ○  professional legal issues (employment law, personal liability issues)
  - ○  future trends in biomedical engineering
  - ○  management and leadership
  - ○  decision making

Few faculty members, if any, are familiar enough with all of these topics to provide up-to-date, relevant, treatments of every topic. In this case, other faculty or working engineers with expertise in specific knowledge areas can be asked to be guest speakers. This helps spread the load among more than just one instructor, provides students with good role models (especially if guest speakers represent a diverse group of experts), and ensures that students are presented with the most accurate, up-to-date information from someone who deals with the specific topic on a daily basis.

Results from the 2005 survey[2] of capstone design programs regarding the subjects taught as part of capstone courses are shown in Figure 7.1. The "other" category included more than 75 additional topics. Subjects involving communication, interpersonal, teamwork, and leadership skills were the most frequently taught along with other professional skills. Subjects involving technical skills and issues were included in capstone courses less frequently. This may be because of the emphasis on professional skills by the ABET engineering criteria and a recognition of the importance of these topics to engineering education, successful completion of the capstone design project, and careers after graduation.

## REFERENCES
1. Wilbarger, J., and Howe, S., *Current Practices in Engineering Capstone Education: Further Results From a 2005 Nationwide Survey*, presented at the ASEE/IEEE Frontiers in Education Conference, San Diego, CA, October 2006.
2. Howe, S., and Wilbarger, J., *2005 National Survey of Engineering Capstone Design Courses*, presented at the 2006 ASEE Annual Conference and Exposition, Chicago, IL, June 2006.

•  •  •  •

CHAPTER 8

# Intellectual Property and Confidentiality Issues in Design Projects

While working on capstone projects, students are encouraged to develop unique, novel, creative design solutions to problems. Many of these problems are the specific focus of industry- or faculty-sponsored projects. Student teams often develop creative design solutions that result in the creation of intellectual property. When this occurs, who owns this intellectual property? Is it owned by students, faculty advisors, industry sponsors, the academic institution, or all of these stakeholders? Does the answer to this question depend upon whether students used university resources such as laboratories, test equipment, machine shops, computer facilities, or departmental funding to create the intellectual property? What if an industry sponsor provided resources or funding to complete the project? Can students be required or allowed to sign patent assignment agreements that assign all rights to intellectual property to the academic institution or the sponsoring company?

Many industry-sponsored projects require sharing of confidential company information with students. To protect themselves, companies often require project team members to sign nondisclosure agreements, which prohibit students from disclosing any confidential information related to the project and the company's businesses, products, and technologies that they may learn of while working on the project. Can the university require students to sign these agreements? What is the legal status of students in a senior capstone design course? Are they considered to be employees of the university, covered by the same policies as faculty, or are they given some other legal status?

What is the policy of academic institutions regarding intellectual property resulting from senior design capstone courses and protection of confidential industry information? There is no clear consistent answer to this question. Each academic institution has its own policies regarding these issues.

Gary Brandenburger, D.Sc., of the Washington University School of Engineering and Applied Science in St. Louis, MO, has been studying how academic institutions deal with the issues involving intellectual property in senior design capstone courses offered by engineering schools in the United States. Preliminary findings indicate that the answers to the following questions have an effect on the sophistication of and degree of industry involvement in senior capstone design courses:

1.  Are undergraduate students permitted to sign confidentiality/nondisclosure and intellectual property assignment agreements?
2.  Are there policies and procedures in place to support (or at least not interfere with) executing such agreements and conducting collaborations with industry sponsors?
3.  Are faculty and administration given logistic and legal support and the time required to develop and manage industry collaborations?
4.  Do the policies of the academic institution allow (or at least not prevent) industry sponsors and students to retain all rights to intellectual property?
5.  Does the design course faculty have the motivation and time to develop and apply methods and practices to assure that academic objectives and ethical standards are met when students work in teams on industry-sponsored projects?

Some universities report that they have satisfactorily addressed such issues, including the challenges of intellectual property and industry collaborations, allowing them to offer sophisticated senior capstone design courses. Others have chosen to constrain the way they offer courses to avoid these issues altogether, such as discouraging development of any intellectual property, not involving industry, or not allowing students to work in teams. Some schools believe that such limiting choices severely constrain the depth, breadth, and richness of the capstone design course experience and in extreme cases reduce it to a semester-long homework problem assignment. Others find compromises that provide as rich an experience as possible without industry involvement. Schools have begun to examine intellectual property issues, and at least two universities have formed task forces to study them.

At Marquette University, students are informed before selecting an industry-sponsored project if the sponsor will require signing of nondisclosure and patent assignment agreements. They are told that this will be a requirement for participation on the project team. If they are not comfortable with signing these agreements, then it is recommended that they choose a project that does not require this. Each year, only a few industry sponsors require these agreements as a prerequisite for sponsorship. Undergraduate students are not considered to be university employees and thus could not be required by the university to sign agreements with industry sponsors. The university acts as an uninvolved third party, and the agreement is between the students and the sponsoring company.

Many senior capstone design courses have evolved into sophisticated learning experiences that prepare students for careers in engineering. Some of these produced entrepreneurial engineers who created start-up companies. These courses generate intellectual property for which policies are needed to determine how it will be treated. Truly innovative design solutions tend to be patentable and thus result in the creation of intellectual property. Policies that discourage the creation of intellectual property in senior design courses, either by not protecting the intellectual property rights of students and faculty or the confidentiality of an industrial sponsor's information or by discouraging industry involvement, lower the probability of developing truly innovative design solutions. Because most of our graduates enter the industrial workforce, this situation hinders engineering educators from accomplishing one of our primary goals: preparing students for engineering careers involving the development of open-ended, creative, and innovative solutions to relevant real-world problems.

Discouraging industry involvement in senior design courses prevents students and academic institutions from reaping the benefits of industry-sponsored projects. For example, sponsoring companies can provide additional resources to project teams that are not available otherwise. Students benefit from industry sponsorship through the opportunity to work on real-world problems of importance to industry, exposure to the medical device industry and market, and familiarity with the requirements and constraints (economic, legal, and regulatory) affecting medical device design.

According to the 2005 survey of capstone design courses, intellectual property generated by industry-sponsored capstone design projects is owned by the industry sponsor (64%), the university (45%), and students (44%).[1] Ownership rights are often shared. Sponsors owned all rights at 59% of the institutions responding to the survey. The university and students shared all rights at 38% of the institutions. Often, one to two thirds of intellectual property rights were assigned to students and the university.

## REFERENCE

1. Howe, S., and Wilbarger, J., *2005 National Survey of Engineering Capstone Design Courses*, presented at the 2006 ASEE Annual Conference and Exposition, Chicago, IL, June 2006.

•   •   •   •

# SECTION III
# Enhancing the Capstone Design Experience

CHAPTER 9

# Industry Involvement in Capstone Design Courses

Faculty mentoring of project teams and the use of project advisors from industry provide valuable learning experiences for students. Industry involvement in senior design courses is not limited to project sponsorship and benefits students, faculty, academic institutions, and participating companies.

Representatives from industry can participate in senior design courses as guest lecturers, curriculum advisors, and design project sponsors. As guest lecturers, practicing engineers provide a relevant and practical real-world perspective of their topic, reinforcing its importance to the practice of biomedical engineering. Students (and design faculty) appreciate the up-to-date presentation of the topic provided by guest lecturers from industry. Practicing engineers in industry are often willing to speak to students about topics in their areas of expertise. It provides them with an opportunity to share their experiences with students and is a professional development activity in which many working engineers are interested. Being asked to be a guest speaker is flattering to many engineers because it suggests that they are considered to be experts in the topic of interest. In areas with local medical device companies, finding local engineers to speak, especially alumni, should not be difficult. Otherwise, travel expenses for guest speakers and honoraria can be offered as incentives.

Feedback from engineers working in industry can be very helpful in ensuring that the content and objectives of senior design courses are relevant to the practice of biomedical engineering and are helping to prepare students for careers in biomedical engineering. A periodic review of course objectives, lecture topics, and required deliverables by members of an industrial advisory committee can help fine-tune the course curriculum.

Industry sponsorship of senior design projects is beneficial to the sponsoring companies, the university, and students. Companies benefit by receiving additional technical resources dedicated toward solving a technical problem, at little or no cost. Many senior design courses require project teams to develop functional prototypes and demonstrate that customer (industry sponsor) needs have been met. Typically, a company will be assigned a team of engineering students dedicated to the project for the duration of the course(s). This is beneficial to companies with limited engineering resources and allows them to make progress on lower priority projects without diluting efforts on those of higher priority. Project sponsorship also allows companies to participate in the training of new engineers, advertise their companies on campus, and gain access to a pool of graduating engineers for recruitment.

Academic institutions benefit from industry sponsorship of senior design projects through the building of relationships with industry (which can lead to research collaborations and grants), maintenance of a high-quality senior design course and project experience, and addition of resources available to students to complete their design projects.

Students benefit from industry sponsorship of senior design projects through the opportunity to work on real-world problems of importance to industry, exposure to the medical device industry and market, experience with project management and the product development process, and familiarity with the economic, legal, and regulatory requirements and constraints of medical device design. Positive senior design project experiences can sometimes lead to full-time employment with the sponsoring company after graduation.

The requirements for industry sponsorship of senior design projects vary among academic institutions. Issues regarding the sponsoring company's level of time and personnel commitment to the project team, funding of the project, and access to company resources need to be finalized before the start of the project. For example, at Marquette University, potential sponsors are required to identify a company representative to act as an industry advisor to the project. The industry advisor acts as the company contact for the team, advises students on issues involving customer needs, provides technical expertise and advice, and approves design concepts and prototypes. Faculty advisors are responsible for administrative issues (grading, monitoring progress of teams, dealing with personnel issues, etc.) and providing guidance and technical advice to the team. Industry advisors are required to be available to discuss project requirements, customer needs, and potential design solutions. Communication can be in person or by telephone, e-mail, or fax. The industry advisor can determine the frequency of communication with the team or team representative as well as the need for travel.

At many schools, senior design teams are required to construct and test prototypes to verify that their design solves the sponsor's problem and meets the sponsor's needs. Students typically have access to the university's computer network, libraries, machine shops, and laboratories. Construction of functional prototypes can be costly, and testing of prototypes may require specialized test equipment not available to students. Depending upon the complexity of the design and the requirements of the sponsoring company, some prototypes can be made of parts obtained from local hardware stores and easily assembled in a dormitory room or laboratory. Other prototypes may require access to a machine shop for lathes, mills, drill presses, and others or must be made of materials that require casting, molding, or other processes that might not be available to students in an academic setting. In these situations, industry sponsors are requested to provide the necessary additional resources (prototyping facilities and/or personnel, laboratories, and test equipment) to teams to complete their projects. Some schools require industry sponsors to pay a fee to sponsor projects, with the funds used to cover expenses for prototype development and testing, travel, and other project- and course-related expenses. For example, the Harvey Mudd Engineering Clinic Program requires industry sponsors to pay a minimum $41,000 (2006–2007) fee to "defray the cost incurred by the college in the connection with the establishment and supervision of the program."[1] Other schools ask that sponsors provide the economic and technical resources necessary for the team to complete their projects. In the 2005 survey of capstone design courses, respondents indicated that funding for capstone design projects was obtained from industry sponsors (52%), academic institutions (68%), students (30%), and other sources (25%).[2] For industry-sponsored projects, various levels of funding were reported including none (6%), <$500 (48%), $500–$1000 (9%), $1000–$5000 (16%), and >$5000 (12%), with 2% of the respondents reporting that they received >$40,000 for projects.

My experience with the two-semester senior capstone design course at Marquette University has shown that certain types of projects are well suited for industry sponsorship. For other design programs, the types of suitable projects will vary depending upon the length and required deliverables of the course(s). The following types of projects will allow students to meet course requirements and the needs of industry sponsors:

- Lower priority projects for which the company lacks resources (this is attractive to start-up companies with few technical resources)
- Projects that can be completed in 9 months or less (required for a two-semester design course sequence)
- Projects involving the development of truly new products (this may be difficult to complete in two semesters), improvements to existing products (new features, revised packaging, new materials, etc.), or process improvements.
- Projects requiring the development of new test procedures and the design of specialized test equipment.

Industry-sponsored projects present challenges regarding intellectual property. Policies regarding intellectual property rights and ownership vary among academic institutions. Some companies require students working on their projects to sign nondisclosure and patent assignment agreements as a condition for sponsorship. Other companies will not allow any public disclosure (classroom presentation) of the results of the projects they sponsor. Students must be informed of the policies of the university and sponsor and must be allowed to decline participation on a project if they are not willing to sign agreements with the sponsor. As long as students and industry sponsors are aware of the intellectual property policies affecting their projects before the start of the project, this should not cause a problem later on if the project results in intellectual property.

There are many benefits to involving industry in senior capstone design courses. Experienced engineers (working or retired) can be guest speakers, curriculum advisors, or project sponsors. This type of industry/academia collaboration benefits the university, students, and sponsoring companies.

## REFERENCES

1. http://www.eng.hmc.edu/EngWebsite/Clinic/06-07ClinicHandbook.pdf
2. Howe, S., and Wilbarger, J., *2005 National Survey of Engineering Capstone Design Courses*, presented at the 2006 ASEE Annual Conference and Exposition, Chicago, IL, June 2006.

•   •   •   •

C H A P T E R  10

# Developing Business and Entrepreneurial Literacy

The senior capstone design course is an excellent tool for preparing students for careers in biomedical engineering, especially those interested in employment with medical device companies.

More than half of all graduate biomedical engineers will be employed in industry. In addition to communication, interpersonal, and design skills and knowledge of the product development process and project management, successful careers in the medical device industry require an understanding of how a business functions (marketing/sales, accounting/finance, and operations) and familiarity with the legal, regulatory, and economic constraints affecting medical device design and development. As members and leaders of project teams, engineers need to understand the roles that team members from other functional areas within the organization play on the team and appreciate their contributions to the organization. Engineers do not need to be experts in these other areas, but they must be able to "speak the language" of their fellow team members. This is where knowledge of other business functions such as accounting, finance, marketing, operations, and regulatory affairs can be beneficial. This additional knowledge base constitutes a minimal level of business literacy needed for successful careers in industry. Depending upon the curriculum of a specific biomedical engineering program, students may be able to supplement their engineering education with courses in accounting, finance, and marketing. This would help develop the minimal level of business literacy that I feel all engineers should possess.

During the last several, many undergraduate biomedical engineering programs became interested in entrepreneurship and in developing entrepreneurial biomedical engineers. Some schools offer courses in entrepreneurship, sponsor business plan competitions, and provide resources conducive to entrepreneurship such as special laboratories for "tinkering" and incubator office and laboratory space for start-up medical device companies. Some schools have incorporated entrepreneurship into their senior capstone design courses, have encouraged the formation, and funded the activities of entrepreneurial project teams (e-teams) consisting of students, faculty, and other nonengineering team members.

The question of whether entrepreneurship could and should be taught at the undergraduate level has been raised. Before I present my opinion, I need to define the term *entrepreneur*. One source defines an entrepreneur as "a person who organizes, manages, and assumes responsibility for a business or other enterprise."[1] Another definition is "one who organizes, manages, and assumes the risk of a business or enterprise."[2] According to these definitions, entrepreneurs need organizational and managerial skills and must be willing and able to assume and manage the risk of running a business or enterprise. People interested in starting new medical device companies (entrepreneurs) also need to know how to identify market opportunities, assess market potential, write a business plan, secure funding, protect intellectual property, and make the best use of limited resources (money, equipment, personnel, etc.).

The skills and knowledge needed to be a successful entrepreneur are in addition to the skills and knowledge needed to be a successful engineer in industry. Not all biomedical engineering students are interested in starting new medical device companies and do not need to be able to write a business plan or know how to obtain venture funding. I would argue that very few, if any, undergraduate biomedical engineering students are qualified to start and successfully operate their own medical device companies without assistance from experienced managers, no matter what courses they have taken as undergraduates or what internships they may have participated in. To suggest the opposite underestimates what it takes to not only start but to grow and maintain a successful medical device start-up company.

In my opinion, as educators and design instructors, our goal should be to provide our students with the opportunity to develop the skills and acquire the knowledge needed for successful careers in biomedical engineering. This includes development of 1) communication, interpersonal, teamwork, and design skills; 2) an understanding of the legal, regulatory, and economic aspects of health care delivery and the medical industry; and 3) basic business literacy. A course or courses that would expose students to the basics of marketing, finance, and how a business operates would be beneficial. For those students who are interested in someday starting a medical device company, their entrepreneurial interests should be nurtured by providing them with the opportunity to develop basic entrepreneurial literacy. This could include courses beyond those providing basic business literacy, those that address sources of venture funding, how to write a business plan, how to organize and structure a start-up venture, technology transfer (including intellectual property protection and licensing strategies), and management of technical personnel. In my opinion, having students write business plans before teaching them something (formally or informally, through an entire course, seminar, or single lecture) about marketing and finance is putting the cart before the horse.

How can we fit these additional topics in an already crowded engineering curriculum and possibly crowded senior design lecture schedule? Some schools allow students to take business courses as part of an undergraduate entrepreneurship program. Other schools may offer seminars on related topics for those interested in entrepreneurship. Perhaps, these topics and courses are better addressed at the graduate level by MBA, engineering management, or other specialized management degree programs. Another option is to take advantage of "teachable moments" that provide the opportunity for students to focus on the business, rather than the technical, side of their senior design projects and other technologies they learn about in other courses. For example, in a graduate course required by the Healthcare Technologies Management program at Marquette University and the Medical College of Wisconsin, various medical devices and technologies are presented by guest lecturers. The technical aspects of the technologies are discussed, but emphasis is placed on the legal, regulatory, and economic aspects. The cost of the device itself and the cost of procedures using the device are presented, along with information on market size and market leaders and their respective market shares. Trends in a particular market are often presented along with potential market opportunities. Issues regarding insurance reimbursement, recalls, regulatory requirements, and outcomes assessment are also discussed. This type of information can be presented as part of any course involving medical devices and technologies. Asking students to think about new applications and potential market opportunities for new or existing medical devices and technologies is a first step in developing and nurturing entrepreneurship in undergraduate biomedical engineering students.

I feel that although developing entrepreneurial biomedical engineers is a worthwhile goal, our focus as design instructors should first be on preparing them for careers in biomedical engineering as engineers. We should do what we can to enable them to become the best engineers possible, including helping them develop a minimal level of business literacy. For those students interested in entrepreneurship, we should provide opportunities to develop and nurture their interest and help them develop a minimal level of entrepreneurial literacy. For additional in-depth entrepreneurial education, a graduate management degree with a concentration in entrepreneurship might be the best approach. To develop an entrepreneurial perspective in all biomedical engineering students, we can take advantage of "teachable moments" that present themselves in many of our courses.

## REFERENCES

1. *The Random House College Dictionary*, Random House, New York, 1973.
2. *Merriam-Webster's Collegiate Dictionary*, 10th ed, Merriam-Webster, Springfield, MA, 2002.

• • • •

CHAPTER 11

# Providing Students With a Clinical Perspective

Successful careers in biomedical engineering require a minimal level of technical, business, and legal literacy. Technical literacy includes knowledge of the engineering sciences, life sciences, mathematics, and computer literacy. Business literacy includes knowledge of basic business functions such as marketing, sales, and finance. Legal literacy includes a familiarity with intellectual property, professional liability, and ethical issues, as well as domestic and international quality and regulatory requirements for medical device design, development, and commercialization.

Some components of technical, business, and legal literacy can be presented to students via lectures included in the senior design course. Additional components of business and legal literacy can be presented in greater depth through other required and elective courses in the engineering curriculum.

Biomedical engineers also need a minimal level of clinical literacy. This includes knowledge of medical and surgical terminology and procedures. It is not only important for students to understand how medical devices are designed and function but how they are used by patients, physicians, surgeons, nurses, physical therapists, and caregivers. This information is required to enable biomedical engineers to design medical devices that truly meet customer (patient, physician, or other end-user) needs and solve relevant medical problems in a cost-effective manner. Clinical literacy includes an awareness of current clinical issues and problems whose solutions may involve the application of technology.

There are several ways of exposing senior design students to the clinical aspects of medical device design. This may be done through courses taken previously or through activities included as part of the senior design course. Lectures and presentations aimed at making students aware of current problems in medicine and how they are being solved through the use of technology are very beneficial to students. However, direct observation of medical procedures involving medical devices combined with student interaction with and questioning of medical professionals can be more beneficial. Students probably learn the most about the clinical aspects of medical technology when allowed to observe and speak with medical professionals and patients directly (subject to appropriate informed consent guidelines and HIPAA regulations to protect patient rights and privacy). This allows them to understand the physician's and patient's perspectives.

There are several ways to provide students with the clinical perspective needed to better understand the requirements of their senior design projects. First, students can observe surgical procedures in an operating room and/or accompany physicians on rounds in a hospital or clinic. Many biomedical engineering programs collaborate closely with nearby medical schools and/or hospitals. Senior design course instructors can set up programs that allow biomedical engineering students to observe various surgical procedures and accompany physicians and surgeons on rounds and office visits to observe how technology is used in the clinical setting. The advantage of direct observation is that students witness firsthand how technology is actually used and potentially misused. Sometimes, medical devices are not always used in ways they were intended, either intentionally or because the devices are too difficult or confusing to use per the manufacturer's recommendations. Direct observation in the clinical setting helps students appreciate the importance of human factors in the design of medical devices, especially user interfaces.

Second, course instructors can obtain video recordings of surgical procedures from medical schools or medical device manufacturers and allow students to view them during or outside of class. By viewing live or recorded surgical procedures, students learn how surgical instruments are used and how implantable devices such as total joint replacements are handled in the operating room, prepared for implantation, inserted into the patient's body, attached to various tissues, and removed. This experience gives students an appreciation for the real problems associated with the use of medical devices.

Third, when visiting hospitals and clinics to observe procedures, if students are allowed to speak with and ask questions of patients, physicians, physical therapists, and other medical personnel, they will gain additional understanding of the real needs and problems that exist with the use of technology in medicine. By speaking with medical professionals, students can learn about the success rates of various devices and procedures, state-of-the-art treatments and technologies used in health care, and real-world problems currently in need of technical solutions. This knowledge can be helpful to students planning careers in the medical device industry and can assist them in identifying problems on which to base their senior design projects.

Finally, students can learn much about problems and issues related to the use of technology in the clinical environment from clinical engineers. Clinical engineers possess a wealth of experience and knowledge regarding the assessment, implementation, use, and maintenance of hospital-based health care technologies such as imaging systems, monitoring equipment, therapeutic devices, and surgical instrumentation. They can provide students with insight into problems with current devices. This is the first step toward helping students identify potential design solutions in the form of new device features and design changes, resulting in lower costs, improved serviceability and maintenance, increased patient safety, and improved usability.

These approaches toward providing students with a clinical perspective can produce additional beneficial outcomes. First, they can help students develop skills in problem identification and determination of customer needs through direct observation of procedures and attentive listening to customers discussing problems for which technical solutions are needed. The ability to observe procedures, listen to customers, and identify problems and customer needs is vital to recognizing potential market opportunities and highly valued by medical device companies. Second, interviewing and discussing potential design solutions with medical personnel helps students learn to communicate technical issues to nontechnical people and work with people in nontechnical disciplines. This is helpful to students who will eventually work on multidisciplinary project teams in industry.

It is not only important for students to understand how medical devices are designed and function but how they are used by patients, physicians, surgeons, nurses, and physical therapists, too. For implantable devices, this includes problems associated with the preparation, insertion, and removal of devices in a patient's body. For monitoring or imaging equipment, this includes problems associated with device setup and operation and interpretation of data. This knowledge is important to medical device designers who may be able to avoid or solve many of these problems through improved design.

•   •   •   •

CHAPTER 12

# Service Learning Opportunities

Many employers consider skills and knowledge related to the social sciences (communication, team-work, and ethics) to be more important than those related to engineering or the natural sciences to the successful performance of new engineers.[1] Six of the 11 outcomes required of engineering programs by the ABET are related to the social sciences (economic and political issues, cultures, societies, and communication), and include the following:

- ability to function on multidisciplinary teams
- understanding of professional and ethical responsibility
- ability to communicate effectively
- broad education necessary to understand the impact of engineering solutions in a global and societal context
- recognition of the need for and an ability to engage in lifelong learning
- knowledge of contemporary issues

To produce these required outcomes, engineering students need knowledge of the social sciences. The senior design project experience can serve as a way to link engineering with the social sciences, and produce the required ABET outcomes.

Most senior design projects have one of four slightly different goals. These goals include 1) the development of equipment used to solve a research-related problem; 2) cost reduction, new-feature implementation, enhancement of an existing device, or development of a new product; 3) solution of a problem specific to an individual patient or group or patients with a similar medical condition; or 4) solution of a problem encountered in the local, regional, national, or global community. The latter is the goal of service learning projects. Often, these projects involve problems experienced in an underdeveloped part of the world.

Service learning can be defined as "action and reflection integrated with the academic curriculum to enhance student learning and to meet community needs."[2] It exposes students to ethical situations and provides community interaction that stimulates reflection, helping students contemplate engineering solutions in a global and societal context.[3] Students involved in service learning perceive an increase in their personal development, tolerance, and communication skills. They are also more motivated to understand technical information and devote more time and thought toward the technical and social tasks at hand.[4] Service learning senior design projects allow biomedical engineering students to work on real-world engineering problems, help people in the local or global community, learn about other cultures, and see how their education and skills can be used to benefit others. These project experiences can help produce the ABET outcomes previously discussed.

Some engineering programs include service learning activities. For example, Purdue University's Engineering Projects in Community Service program brings together engineering students and community agencies to develop technical solutions to benefit the community. Examples include energy management systems for Habitat for Humanity, database systems for homelessness prevention organizations, and speech recognition software for speech and audiology clinics.[5]

At Marquette University, the Health, Environment, and Infrastructure in Latin America plan was initiated to promote undergraduate engineering experiences in international service learning. The plan includes identification of international service locations, development of multidisciplinary courses for students participating in international service learning projects, and coordination of international projects within existing course work. Students travel to service locations to work on infrastructure and health care projects. Project teams can consist of biomedical, civil, environmental, mechanical, and electrical engineering students. Service locations include El Salvador (health care clinic), Haiti (low-cost renewable power source for grade school), and Guatemala (bridge construction and sanitary sewer). The sanitary sewer project in Guatemala was completed as part of the student team's senior capstone design course.[6]

A few years ago, service learning projects were included in the biomedical engineering senior capstone design course at Marquette University. For example, a project team of engineering students sponsored by the Department of Physical Therapy developed a low-cost system for assessing lung function in Latin American garment workers. Students on the team were given the opportunity to travel to Latin America to assist in testing of pulmonary function of garment workers using the device they designed. The following year, another team of biomedical and mechanical engineering students developed a low-cost (<U.S.$50), low-maintenance wheelchair for disabled people in El Salvador. Several more international service learning projects have been completed during the last few years.

A potential source of international service learning projects for biomedical engineering senior design students is Engineering World Health. This organization is dedicated to the biomedical needs of the developing world. They provide international service learning opportunities for biomedical engineering students. Medical equipment and parts are donated, and volunteers use their time and expertise to restore the equipment for reuse. The refurbished equipment is then delivered to and installed in a community in need.

Engineering World Health sponsors a *Design Projects that Matter* (designpprojects@ewh.org) series through which students can obtain funding for senior design projects resulting in designs that meet specifications determined by the organization. The objective of these projects is to design test equipment that can be freely distributed to the developing world. If a team's design is selected, funds are made available to produce a prototype. If the prototype functions as required, it may be selected for production, and the team will be given the opportunity to travel to a developing nation to deliver and implement the new equipment.

There are many benefits to incorporating service learning projects into senior capstone design courses. The experience can be life-changing for students because they realize the impact that they can have on health care delivery in the developing world and understand the importance and effects of engineering solutions on the world and society.

# REFERENCES

1. Valenti, M., "Teaching tomorrow's engineers," *Mechanical Engineering (American Society of Mechanical Engineering)*, vol. 118, pp. 64–69.
2. O'Grady, C., "Integrating service learning and multicultural education: an overview," in *Integrating Service Learning and Multicultural Learning in Colleges and Universities*, Lawrence Erlbaum Associates, Mahwah, NJ, pp. 221–232, 2000.
3. Haws, D.R., "Ethics instruction in engineering: A mini meta-analysis," *Journal of Engineering Education*, vol. 90, pp. 223–230, 2001.
4. Eyler, J., Giles, D.E., and Braxton, J., "The impact of service learning on college students." *Michigan Journal of Community Service Learning*, vol. 4, pp. 5–15, 1997.
5. Coyle, E.J., Jamieson, L.H., and Sommers, L.S., "EPICS: a model for integrating service learning into the engineering curriculum," *Michigan Journal of Community Service Learning*, vol. 4, pp. 81–89, 1997.
6. Zitomer, D.H., and Johnson, P., "International service learning in environmental engineering," *World Water and Environmental Resources Congress 2003*, P. Bizier and P. DeBarry, editors, June 23–26, 2003, Philadelphia, PA.

• • • •

CHAPTER 13

# Collaboration With Industrial Design Students

When biomedical engineering graduates enter the workforce, they will be expected to work on multidisciplinary teams. In industry, these teams typically consist of members of research and development, marketing, production, finance, regulatory affairs, and other departments. Depending upon the type of products being developed, customer needs, and specific design requirements, industrial designers may be assigned to the project team to work with engineers on the design of the new product. Industrial designers are uniquely qualified to assist with specific aspects of product design. However, many biomedical engineering students and faculty members are not aware of what industrial designers actually do and the role that they can play in the development of medical devices. To prepare biomedical engineering and industrial design students for potential future collaborations, it would be helpful for them to understand and appreciate the contributions each can make to the project team. This can be accomplished through the senior capstone design course by forming project teams that include biomedical engineering and industrial design students.

To appreciate the role of industrial designers in the design process, it is helpful to understand the three main aspects of medical device design.[1] First, the technical aspects involve the assembly of parts and systems that allow the device to meet the technical requirements. Second, the human factors aspects deal with how well the user interface enables the user to interact with the device, encourages correct performance, and discourages and prevents incorrect performance. Third, aesthetic form can communicate how to use a device to achieve the intended result and can make a product easy to use. Although the appearance of a device has little effect on its user interface, it can have a strong psychological influence on the patient or end user. All three aspects of design help create value and enhance the overall perception of quality. A well-designed medical device satisfies all customer needs, meets all required specifications, incorporates basic human factors principles, and is sensitive to aesthetics and market perception.[1]

Engineers and industrial designers tend to emphasize different aspects of design. Engineering students (and practicing engineers) tend to focus on the technical aspects of design such as functionality and performance specifications. For example, engineers developing implantable devices are concerned with issues such as corrosion, wear, degradation, strength, and fatigue life. They perform calculations, use a variety of analytical tools (such as finite element analysis), construct analytical models, and conduct bench tests on prototypes to ensure that products are made from materials with the appropriate design characteristics (strength, biocompatibility, biodurability, etc.) and will safely perform as required. Industrial designers focus on usability, safety, quality, and the aesthetics of products. They are concerned with issues such as the psychological impact of a product's design on the user or potential customer, usability (ease of use, low potential for error), safety (no sharp edges or other potential hazards), quality of the overall product experience, and perceived value of the product. Engineers are also concerned with these issues, but engineering curricula typically do not spend as much time on aesthetics and usability as do industrial design curricula. Both disciplines place heavy emphasis on identification of customer needs, manufacturing methods, and prototyping.

During a tour of the Milwaukee Institute of Art and Design (MIAD; Milwaukee, WI), I was impressed by the quality of the models on display and the model-making resources available to the industrial design students. Most of the models were not functional but were very professional looking. Historically, the prototypes developed by the students in my senior design classes have been functional but not aesthetically pleasing. After seeing what the industrial design students could produce, I was convinced that including industrial design students from MIAD in the project teams with biomedical engineering students from Marquette University would improve the quality of the projects and result in prototypes that were aesthetically pleasing and functional.

To encourage collaboration between industrial design and biomedical engineering students, Pascal Malassigne, FIDSA, professor of industrial design at MIAD, created a course that would give credit to his students for working on senior design projects with students at Marquette University. Because of scheduling issues, the MIAD students would only be available to work on the Marquette projects for part of the second semester of the two-semester senior design course. At the start of the spring semester, three groups of three MIAD students were assigned to three of the Marquette projects currently in progress. The MIAD students functioned as design consultants to the project team and used their model shop to produce prototypes. This collaboration produced professional-quality, aesthetically pleasing, functional prototypes. When surveyed about this collaboration, the MIAD and Marquette students said that they had learned about the other's discipline and the role each played on the project team. They developed an appreciation for the other's contributions to the team and felt better prepared to work with each other on product development teams in the medical device industry.

At the University of Cincinnati, Mary Beth Privitera, M.Design, assistant professor in the Department of Biomedical Engineering and the Medical Device Innovation and Entrepreneurship Program, teaches a course that teams business, industrial design, and biomedical engineering students with a physician to study a particular device, learn how it is used, and determine how it could be improved.[2] Each student brings his or her unique skills and knowledge to the project team. The business students identify stakeholders and determine regulatory status, the industrial design students conduct task analyses, and the biomedical engineering students analyze the device and determine how it functions. This course provides students with the opportunity to work on multi-functional teams and develop "cross-language skills" needed for careers in medical device product development. Biomedical engineering students complete this course before enrolling in the required senior capstone design course. The business and industrial design students are invited to continue their multifunctional team experience via participation in senior capstone design projects.

Faculty involved in collaborative design project experiences from the Medical Device Innovation and Entrepreneurship Program at the University of Cincinnati have made some interesting observations concerning transdisciplinary learning among students in different disciplines[3]:

- Biomedical engineering students were familiar with the legal and regulatory requirements for detailed record keeping of project activities and decisions. However, industrial design students were unfamiliar with this practice. This presented a challenge because they were encouraged to record and document their activities.
- Engineers are perceived as thinking in a more linear and causally linked form as opposed to the more lateral or freethinking style of industrial designers. An appreciation for the merits of both styles of thinking was necessary for all team members to feel that they were successful contributors.
- The recognition of the value that each discipline brings to the project team was an essential component of effective transdisciplinary learning. During technical design review meetings where design progress was presented to faculty, engineering students learn to value the industrial design students' ability to communicate complex procedural diagrams coupled with new device concept drawings, and the industrial design students learned to value the engineering students' ability to conduct, analyze, and present test data to prove the technical and clinical advantages of different designs. Students developed an appreciation of each other's complementary functional strengths.
- Collaboration between biomedical engineering and industrial design students on senior design project teams provides many benefits. First, students learn how to communicate with people in other functional disciplines. Second, students learn that no individual person has all the skills and knowledge needed to complete a project, and they develop an appreciation for the complementary skills each member brings to the project. Third, students learn that there is more than one way to solve a problem. This helps them develop an appreciation for different approaches to problem solving and ways of thinking. Finally, the overall quality of product design increases when biomedical engineering and industrial design students work together.

## REFERENCES

1. Hyman, W.A., and Privitera, M.B., "Looking good matters for devices, too," *Medical Device and Diagnostic Industry*, vol. 27, pp. 54–63, May 2005.

2. Beckman, W., "Profile: Mary Beth Privitera, Master of Design," University of Cincinnati, Cincinnati, OH, September 2004, http://www.uc.edu/profiles/profile.asp?id=1913.

3. Privitera, M.B., and Zirger, B.J., "Letting the grain out of the silo: Transdisciplinary product development education," *Innovation, Quarterly Publication of the Industrial Designers Society of America*, pp. 49–51, Winter 2006.

• • • •

CHAPTER 14

# National Student Design Competitions

Several national design competitions exist for engineering students in disciplines other than biomedical engineering. The National Concrete Canoe Competition (American Society of Civil Engineers), Human Powered Vehicle Competition (American Society of Mechanical Engineers), Formula Race Car (Society of Automotive Engineers), and other design competitions are targeted toward civil, mechanical, and other engineering students. In these competitions, student teams representing their respective academic institutions compete against each other. Students are very enthusiastic about these competitions and take them very seriously. They spend many hours developing design solutions and take pride in their work. These competitions provide an incentive and opportunity for students to learn about engineering design and teamwork, and they have been very successful over the years.

Within the biomedical engineering community, there is an interest in providing biomedical engineering students with the same incentives and opportunities to learn about engineering design and teamwork. Annual national design competitions involving the solution of a medically related problem are of interest to biomedical engineering educators, medical device companies, and students. Design innovations and intellectual property generated as a result of the competition are of interest to the medical device industry and investment community and have been the impetus for the formation of new business ventures and start-up companies. A national biomedical engineering design competition can serve as a way of bringing the academic and industrial biomedical engineering communities together, enabling the sharing of best design education practices and dissemination of effective learning tools.

National student design competitions emphasizing technical solutions to medical problems allow students and faculty to showcase the results of senior capstone design projects. They provide a way for academic institutions to publicize their design programs and help medical device companies become familiar with the quality and content of capstone design projects. They are potential sources of capstone design project ideas and funding.

There are several national student design competitions of interest to the biomedical engineering community. Some emphasize entrepreneurship and potential for commercialization and provide awards in the form of cash to be used for further development of the design concept. Others focus more on technical aspects and how well the design meets the needs of a specific patient population. Submissions to these competitions are not limited to capstone design projects or projects associated with any specific course. However, the requirements of many senior capstone design courses result in the creation of many of the deliverables required for submission to design competitions. These deliverables are of sufficient quality to be highly competitive with other entries in the competitions. With very little additional work (if any), project teams can enter these competitions. Team membership is not limited to biomedical engineering students. Entries from multidisciplinary teams are strongly encouraged.

## 14.1 BIOMEDICAL ENGINEERING INNOVATION, DESIGN, AND ENTREPRENEURSHIP AWARD

*Sponsors: National Collegiate Inventors and Innovators Alliance, National Science Foundation, Canon Communications, Industrial Design Society of America, Biomedical Engineering Society, and the Council of Chairs of Bioengineering and Biomedical Engineering Programs*

This competition seeks student design projects that focus on health-related technology and meet a relevant clinical need. Teams must submit a project description, including documentation of the final design, proof of functionality, assessment of patentability (patent search and search for prior art), anticipated regulatory pathway, estimated production costs, and a business plan with a market analysis and details regarding the strategy for commercialization. Judging criteria include technical, economic, and regulatory feasibility, contribution to health care and quality of life, technical innovation, and potential for commercialization. First-, second-, and third-place awards of $10,000, $2500, and $1000, respectively, are presented at the Medical Design Excellence Awards ceremony during the Medical Device Manufacturing East Convention in the presence of many of the top medical device designers and manufacturers. This competition provides students with the opportunity to win substantial product development funding, gain exposure to the medical device industry, and develop industry contacts. More information is available at www.nciia.org.

## 14.2 REHABILITATION ENGINEERING RESEARCH CENTER ON ACCESSIBLE MEDICAL INSTRUMENTATION NATIONAL STUDENT DESIGN COMPETITION

*Sponsor: Rehabilitation Engineering Research Center on Accessible Medical Instrumentation*

This competition provides up to $2000 per team for prototype development costs, awards cash prizes of up to $1000, and will pay for travel to present an accepted paper at a major conference.

Three target design areas, each having to do with accessible medical instrumentation, are defined at the beginning of each academic year. Hypothetical clients with various disabilities are described, and project teams are required to develop designs in the target design areas that will be easily used by specific clients. Teams are required to create a website to be used to evaluate their projects. The websites include a final report, detailed photos, and video of the prototype in use to demonstrate function, engineering analyses, an accounting of all project expenses (not to exceed $2000), estimated production costs, and a discussion of how the design addresses the needs of the specific clients. More information is available at www.rerc-ami.org.

## 14.3 ENGINEERING IN MEDICINE AND BIOLOGY SOCIETY UNDERGRADUATE STUDENT DESIGN COMPETITION

*Sponsor: IEEE Engineering in Medicine and Biology Society*

Entry into this competition requires students to design and construct a device that solves a problem in medicine or biology. The device must not be commercially available. However, significant design modifications to existing products are acceptable. Product designs may consist of hardware, software, or a combination. Submissions include a project summary, background of the problem to be solved, purpose and scope of the project, technical description of the design, supporting technical analyses, conclusions, and design specifications. Verification of project success must be supported with videotape and/or photographic evidence. Awards of up to $300 and a plaque are presented at the awards ceremony during the IEEE Engineering in Medicine and Biology Society Annual Conference. Winning teams are provided with travel stipends and conference registration. Winning entries are published in *IEEE Engineering in Medicine and Biology Society Magazine*. More information is available at www.ieee.org.

## 14.4 THE COLLEGIATE INVENTORS COMPETITION

*Sponsors: National Inventors Hall of Fame, U.S. Patent and Trademark Office.*

The goal of this international competition is to create excitement and interest in technology and economic leadership. It encourages college students to combine knowledge of science, engineering, and mathematics with technical skills to develop creative inventions. The competition recognizes the potential for the student–mentor relationship to produce patentable inventions.

Reduced-to-practice ideas or workable models developed by student teams may be entered in this competition. The invention must be original and reproducible. Entries must include a summary of current literature, results of a patent search, test data and analyses to prove functionality, and a discussion of the environmental, societal, and economic benefits of the invention. Judging is based on originality, inventiveness, and workability of the invention; quality of supporting documentation; and the invention's potential value to society. The award for the best undergraduate student or student team is $15,000 for the team, and $5000 for the advisor. More information is available at www.invent.org/collegiate.

## 14.5   NATIONAL SCHOLAR AWARD FOR WORKPLACE INNOVATION AND DESIGN

*Sponsor: NISH*

The purpose of this competition is to encourage students to design creative technical solutions that overcome barriers that prevent people with disabilities from entering or advancing in the workplace. Individual students or student teams are encouraged to submit workplace technology designs related to technology for special populations, computer access and use, environmental accommodations, functional control and assistance, transportation/mobility, service delivery, and augmentative and alternative communication. Awards of up to $10,000 are provided to the best designs that will help create employment opportunities for people with severe disabilities. A matching gift is given to the sponsoring department. Winners are expected to attend the NISH National Training and Achievement Conference and present a poster exhibit and an oral platform presentation during a conference session. Judging is based on relevance of the design to NISH goals, quality of the abstract, background discussion, problem statement, appropriate design and evaluation methods, and results and discussion. More information is available at www.nish.org.

The design competitions presented here welcome entries from biomedical engineering students and multidisciplinary project teams. Several have been established to encourage students to use their knowledge and skills to solve medical or other related problems. Some encourage design projects in specific areas of engineering, whereas others focus more on entrepreneurship, innovation, and commercialization. They are potential sources of capstone design project ideas and funding. As engineering educators, we should encourage our senior design students to enter these competitions. Perhaps, the next major innovation in medical technology will come from a project that was a winner of one of these competitions.

•  •  •  •

CHAPTER 15

# Organizational Support for
# Senior Capstone Design Courses

The goal of any engineering program should be to prepare students for their professional and personal lives after graduation. For biomedical engineers, this could include a career in biomedical engineering, medicine, law, business, or other discipline. Because most of our graduates eventually work in industry at some point in their careers, the goal of biomedical engineering programs should be to produce graduates who are ready to function as biomedical engineers in industry with little additional training. Graduates should 1) know how to develop and commercialize a new product, 2) understand the technical, economic, legal, regulatory, and environmental constraints of medical device design, 3) effectively function on a multidisciplinary team, and 4) apply the necessary skills and knowledge toward the design and development of solutions that meet customer needs.

The senior capstone design course is the most important component of an engineering design program and plays an important role in the creation of biomedical engineers who begin their careers with the skills and knowledge needed to be successful in a short period.

To meet the learning objectives of the senior capstone design course, support from the biomedical engineering department (chair and faculty), the college of engineering administration, and industry partners is necessary.

A successful biomedical engineering senior capstone design course requires a dedicated instructor who understands and has experience with engineering design, project management, and the product development process used in the medical device industry today. Ideally, the instructor would have some amount of experience in the medical device industry to share with students and use to guide course design and teaching activities.

In capstone courses involving many projects and teams, one instructor cannot effectively advise all project teams. In this case, additional faculty members with the required technical expertise for a particular project are needed to volunteer to advise project teams. For this arrangement to be effective, the department chair and faculty project advisors must 1) recognize the importance of the capstone design experience in helping prepare students for careers in biomedical engineering, 2) support the goals of the course, 3) become familiar with the medical device design and product development process, and 4) be willing to commit their time to advising student project teams. In some departments where it is assumed that all faculty members will advise a project team, faculty members consider project advising as part of their jobs, and volunteer advisors are easy to find. In other departments, the department chair might need to provide incentives, such as additional salary or credit toward a reduced teaching load to get faculty members to participate. Junior faculty may not want to spend time advising project teams if this activity does not count toward the requirements for tenure. In this case, allowing time spent advising project teams to count toward the requirements for tenure could help get more junior faculty involved. In addition to providing the human resources needed to advise project teams, departments should provide the financial resources needed to manage the senior capstone design course. This could include salaries for support staff, honoraria for guest speakers, and money for teams to build and test prototypes, especially for teams without industry sponsors who would normally provide resources for these activities. Design validation activities for biomedical engineering projects require facilities for bench testing as well as access to patients/users. Human clinical studies are beyond the scope of senior capstone design courses because of time and funding requirements. Cadaver testing can be used to quickly validate designs for some projects and requires no regulatory or institutional approvals. Providing teams with access to a cadaver laboratory would help expedite projects. Creation and maintenance of a library containing up-to-date industry standards would also be helpful to student design projects.

The college of engineering needs to provide support that the departments may not be able to provide. For example, additional funds could be made available for project teams as well as the resources needed to construct and test prototypes such as a machine shop, prototyping facilities, and test equipment. The dean of the college must be supportive of the goals of the design program and the capstone design course and must find ways to reward faculty for participating in the course as project advisors. For capstone design courses involving students from several engineering disciplines (biomedical, electrical, mechanical, etc.), the college of engineering should also provide adequate staff for managing the course as well as funding for guest speaker honoraria and collegewide design competitions. The university administration can support design projects by providing a responsive institutional review board to expedite projects involving human testing and establishing a faculty- and industry-friendly intellectual property policy.

Support from other programs in other colleges within the university is needed for senior capstone design courses in which entrepreneurship is an important component. To create truly multidisciplinary entrepreneurial project teams consisting of students from other colleges such as business, arts and sciences, law, fine arts, dentistry, medicine, physical therapy, or nursing, incentives to encourage students (and faculty) from these different programs to work together on projects must be provided. Requiring participation or offering elective credit for participation in the senior capstone design course can help get more students from other disciplines involved in design projects.

Support from industry partners is needed to maintain a successful senior capstone design course. Support can be in the form of guest speakers, project sponsors and advisors, and curriculum advisors. Guest speakers currently working for medical device companies can provide up-to-date treatments of various topics of importance to the medical device industry. Engineers from medical device companies can sponsor design projects and advise project teams on technical and other issues. Industry sponsorship typically includes providing technical advice and the resources needed to build and test prototypes. Working engineers can advise biomedical engineering departments on the components of the senior capstone design course as well as the entire biomedical engineering curriculum, helping to ensure that they remain relevant and up-to-date.

Creating a culture of design throughout all 4 years of the biomedical engineering curriculum is an excellent way to prepare students for a successful capstone design experience and support the course. This can be done by modifying some existing courses to provide additional opportunities for students to solve open-ended problems and develop their communication, interpersonal, and design skills in courses throughout the 4-year curriculum instead of only the senior capstone design course. Course modifications could include adding assignments or classroom activities that would require students to 1) complete group projects, 2) speak in front of groups, 3) write technical reports and other documents, 4) complete design projects (team or individual), and 5) compete in design competitions contained within a particular course. If only one of each of these modifications is made to each of a few courses in the biomedical engineering curriculum, students would benefit from the additional opportunities to further develop their skills and would be better prepared for the senior capstone course and beyond. By including a design project in courses offered during the freshman, sophomore, and junior years, biomedical engineering design programs can create a 4-year design curriculum. Sponsoring departmental or college of engineering design competitions, creating a senior design project public showcase event, and providing resources for non-project–related design activities (that would allow students to "tinker" with their own ideas not associated with any course or course related project) are additional ways to help create a culture of design.

A senior capstone biomedical engineering design course needs support from the biomedical engineering department chair and faculty, the college of engineering administration, other colleges within the university, and industry partners to help produce students who are ready to be productive engineers in industry. This support includes leadership, intradepartmental and interdepartmental cooperation, project team mentorship, faculty incentives, financial and technical resources, and the creation of a culture of design within the engineering curriculum.

• • • •

# SECTION IV

# Meeting the Changing Needs of Future Engineers

CHAPTER 16

# Capstone Design Courses and the Engineer of 2020

By 2020, many demographic, political, social, and environmental changes are expected. Scenarios involving these changes were presented in a 2004 report by the National Academy of Engineering (*The Engineer of 2020: Visions of Engineering in the New Century*).[1] These changes will create new challenges for the engineering profession, which will require updating and revising engineering curricula and teaching practices to ensure that engineering graduates and practicing engineers will be prepared to meet the future challenges and needs of society. The authors[1] of the report presented their vision of the engineer of 2020:

> We aspire to engineers in 2020 who will remain well grounded in the basics of mathematics and science, and who will expand their vision of design through a solid modeling in the humanities, social sciences, and economics. Emphasis on the creative process will allow more effective leadership in the development and application of next-generation technologies to problems of the future.

> We aspire to an engineering profession that will rapidly embrace the potentialities offered by creativity, invention, and cross-disciplinary fertilization to create and accommodate new fields of endeavor, including those that require openness to interdisciplinary efforts with non-engineering disciplines such as science, social science, and business.

> By 2020, we aspire to engineers who will assume leadership positions from which they can serve as positive influences in the making of public policy and in the administration of government and industry.

> Is it our aspiration that engineers will continue to be leaders in the movement toward use of wise, informed, and economical sustainable development. This should begin

in our educational institutions and be founded in the basic tenets of the engineering profession and its actions.

We aspire to a future where engineers are prepared to adapt to changes in global forces and trends and to ethically assist the world in creating a balance in the standard of living for developing and developed countries alike.

The authors[1] also presented their vision for engineering education in 2020:

It is our aspiration that engineering educators and practicing engineers together undertake a proactive effort to prepare engineering education to address the technology and societal challenges and opportunities of the future. With appropriate thought and consideration, and using new strategic planning tools, we should reconstitute engineering curricula and related educational programs to prepare today's engineers for the careers of the future, with due recognition of the rapid pace of change in the world and its intrinsic lack of predictability.

Our aspiration is to shape the engineering curriculum for 2020 so as to be responsive to the disparate learning styles of different student populations and attractive for all those seeking a full and well-rounded education that prepares a person for a creative and productive life and positions of leadership.

The authors presented a list of the required attributes of the engineer of 2020 assuming that 1) technological innovation continues to change rapidly; 2) the world will be intensely globally interconnected; 3) the people who will be involved with or affected by technology (designers, manufacturers, distributors, and users) will be increasingly diverse and multidisciplinary; 4) social, cultural, political, and economic forces will continue to shape and affect the success of technological innovation; and 5) the presence of technology in our everyday lives will be seamless, transparent, and more significant.[1] These include strong analytical skills, practical ingenuity, creativity, communication skills, mastery of business and management principles, leadership skills, high ethical standards, strong sense of professionalism, dynamism, agility, resilience, and flexibility. Engineers must also be lifelong learners.

Many of these attributes are identical to those required of today's engineers. Most are addressed by current biomedical engineering programs and senior capstone design courses. Some programs are addressing these expected changes now. For example, at a recent conference on capstone design courses, more than one program reported that students were working on multidisciplinary virtual project teams with students from other universities. Through this experience, students learn about three current and future trends in product development: globalization, the use of virtual project teams, and working with people from different social, political, and cultural backgrounds. As demographic, political, social, and environmental changes occur, we will constantly need to revise, modify, and update our capstone design courses and possibly our teaching styles to ensure that we continue to provide our students with the tools and opportunities to develop the skills they will need to meet the new challenges of the future.

# REFERENCE

1. "The Engineer of 2020: Visions of Engineering in the New Century," *National Academy of Engineering*, National Academies Press, Washington, DC, 2004.

•  •  •  •

# Conclusion

The biomedical engineering senior capstone design course is probably the most important course taken by undergraduate biomedical engineering students. It provides them with the opportunity to apply what they have learned in previous years, develop their communication (written, oral, and graphical), interpersonal (teamwork, conflict management, and negotiation), project management, and design skills and learn about the product development process. It also provides students with an understanding of the economic, financial, legal, and regulatory aspects of the design, development, and commercialization of medical technology.

The capstone design experience can change the way engineering students think about technology, themselves, society, and the world around them. It gives them a short preview of what it will be like to work as an engineer. It can make them aware of their potential to make a positive contribution to health care throughout the world and generate excitement for and pride in the engineering profession.

Working on teams helps students develop an appreciation for the many ways that team members with different educational, political, ethnic, social, cultural, and religious backgrounds look at problems. They learn to value diversity and become more willing to listen to different opinions and perspectives. Finally, they learn to value the contributions of nontechnical members of multidisciplinary project teams.

"It takes a village" to develop and maintain a properly designed capstone design course. One person cannot do it alone. To maximize the value and effect of the senior capstone design course, support is needed from faculty, department chairs, deans of engineering schools, university administration, and industry partners. In addition to technical, communication, and interpersonal skills, successful careers in biomedical engineering require a minimal level of clinical, business, entrepreneurial, and legal literacy. The senior capstone design course can be used to help develop these literacies.

In summary, the capstone design course plays a critical role in preparing students for their professional and personal lives after graduation. A properly designed course can help produce graduates who 1) are ready to function as biomedical engineers in industry with little additional training, 2) know how to develop and commercialize a new product, 3) understand the technical, economic, legal, regulatory, and environmental constraints of medical device design, 4) are able to work on multidisciplinary teams, and 5) can apply the necessary skills and knowledge toward the design and development of solutions that meet customer needs.

As with any new product, good product development practice requires frequent reevaluation of customer needs and implementation of any design changes needed to meet any new needs. The same should be done with senior capstone design courses as demographic, political, social, and environmental changes make it necessary to revise the objectives, content, and design of the courses. This will help ensure that capstone design courses continue to prepare our students for future challenges and successful, meaningful careers in engineering.

• • • •

# Author Biography

**Jay R. Goldberg** is an associate professor of biomedical engineering at Marquette University and director of the Healthcare Technologies Management program at Marquette University and the Medical College of Wisconsin (Milwaukee). He teaches courses involving project management, new product development, and medical device design. His experience includes development of new products in urology, orthopedics, gastrointestinal, and dentistry. He is a licensed professional engineer in Illinois and Wisconsin. Dr. Goldberg earned a Bachelor of Science degree in general engineering from the University of Illinois and a Master of Science degree in bioengineering from the University of Michigan. He has a Master of Science degree in engineering management and a Doctor of Philosophy degree in biomedical engineering from Northwestern University. He holds six patents for urological medical devices. Dr. Goldberg also serves as chairman of the Subcommittee on Urological Devices and Materials of the American Society for Testing and Materials. Before moving into academia, he was director of technology and quality assurance for Milestone Scientific Inc. (Deerfield, IL), a start-up dental product company. Before that, he worked for Surgitek (Racine, WI), Baxter (Deerfield, IL), and DePuy (Warsaw, IN). He is a member of the Biomedical Engineering Society, the National Society of Professional Engineers, and the Association for the Advancement of Medical Instrumentation and a consultant to the Gastroenterology and Urology Therapy Device Panel of the Medical Device Advisory Committee of the Food and Drug Administration. Dr. Goldberg is a cocreator of the Biomedical Engineering Innovation, Design, and Entrepreneurship Alliance National Student Design Competition and writes a quarterly column on senior design for *IEEE-EMBS* magazine.

Printed in the United States
by Baker & Taylor Publisher Services